junior Charles Catton

Animals Drawn from Nature and Engraved in Aquatinta

junior Charles Catton

Animals Drawn from Nature and Engraved in Aquatinta

ISBN/EAN: 9783744739993

Printed in Europe, USA, Canada, Australia, Japan

Cover: Foto ©berggeist007 / pixelio.de

More available books at **www.hansebooks.com**

ANIMALS

DRAWN FROM NATURE,

AND

ENGRAVED IN AQUA-TINTA,

BY

CHARLES CATTON, Jun.

WITH A DESCRIPTION OF EACH ANIMAL.

LONDON:

Printed for the AUTHOR, No. 7, on the Terrace, Tottenham-Court Road; and sold by I. and J. TAYLOR, at the Architectural Library, No. 56, Holborn, opposite Turn-stile.

M.DCC.LXXXVIII.

TABLE OF CONTENTS,

AND ORDER IN WHICH TO ARRANGE THE PLATES.

THE LION

London, Published by ... ...

by Longman & Co. February 1, 1807

The L I O N,

BEING univerfally efteemed king of the foreft, deferves a pre-eminence of ftation: we therefore introduce him firft to notice.

The Lion is claffed by the Naturalifts in the Cat Tribe; 'tis an animal, of all others the moft majeftic and ftately; with a large head, the upper part of which, with the chin, the whole of the neck and fhoulders, are clothed with a long fhaggy hair, refembling a mane; the hair of the body and limbs is fhort and fmooth, but long on the bottom part of the belly; a long tail, which appears of equal thicknefs by reafon of the increafing length of hair towards the end, where it terminates in a large black tuft: the colour is tawny, or dirty brown; the belly part inclines to white. The limbs of this animal are of vaft ftrength, and need only be feen to convince of their force. The country where they moft abound is Africa, the moft wild and defert parts: they alfo are found in parts of Afia, but the former appears moft congenial to the Lion's conftitution, thofe of that country being much larger (having been known 5 feet high, and 10 feet long), and more fierce than of any other place; the fiery rays of a torrid climate imparting a feverifh heat, which animates them with an invincible courage. The fmell of the Lion is not fo per-fect as in many other animals, but his fhod of rearing fupplies this defect; for, according to all report, when he roars, he puts his mouth to the ground, thus the found is univerfally diffufed, and not coming ftronger from one place than another, the terrified animals, in their hafte to efcape, frequently run to the very fpot they mean moft to avoid; which, by a kind of bounding, he quickly fecures, ftriking it with great force with his paw: he fometimes invades the flocks, and, with eafe, will carry off a tolerable fized ox; he frequently lies couchant, as ex-preffed by Shakefpeare, "with cat-like watch," and fprings upon his prey by furprife; which if he chances to mifs, in a kind of fhame-faced manner, he meafures back the diftance, ftep by ftep, as if to fee in which he erred;——too much or too little. This animal will fuftain hunger for many days, but requires a more frequent fupply of drink, which it laps like a cat, and at every opportunity.—Mr. Buffon obferves the courage of the Lion diminifhes in proportion as his abode approaches to an inhabited country; his confcioufnefs of man's fuperiority and enmity, awakens him to fear and caution; and the ftately Lion, the defpoiler of thoufands, is frequently levelled with the duft by their addrefs:—Three or four men ufually go to the attack on horfeback; thefe if the Lion difcovers at a fmall diftance, he takes to his heels as faft as he can; if at a fmall diftance only, he then walks off, but in a flow and furly manner, without hurry, as if above fhewing fear; as the hunters approach, he flackens his pace, eying his purfuers afkant; finally mak-ing a full ftop, he turns round to face them, gives himfelf a fhake, and roars with a fhort fharp tone—this is now the time for the attack; he who is moft advantageoufly fituated fires his gun, and gallops off; the Lion immediately purfuing, another then fires, and by thus relieving each other, and repeating their fhot, they rarely fail to reduce the tyrant of the foreft. He is fometimes taken in pitfalls, but more generally when a cub, during the abfence of the dam:—they may be rendered tame and docile, though, at times, fubject to refume their native fiercenefs: their gene-rofity and mercy is not lefs confpicuous than their courage; Pliny reports that "they fpare the proftrate, and, when hungry, feize firft on men rather than women, and never on infants, unlefs preffed by great hunger."

We fhall clofe this account by referring to two remarkable inftances of the memory and generofity of this noble animal which are recorded in the Guardian, No. 139. and No. 146.

LIONESS.

The L I O N E S S.

T H E want of that majestic and graceful ornament the mane, is the principal difference between the male and female Lion; thus unadorned, the female Lion appears scarcely to belong to the same species with the male; this diffimilarity of figure influenced us to present a drawing of the Lionefs; the variations will be beft underftood by comparing their two figures.

The female Lion is, in general, fmaller than the male; but, having the fame habits, and being affuated by like propenfities, we fhall take this opportunity of enlarging the defcription already given of this noble animal. Our choice of information will, we hope, be approved when we affert, it was collected in the midft of their haunts, furrounded by their prefence, and being corroborated by many witneffes, appears to have manifeft advantage over the generality of reports, which ufually have paffed through feveral relators. To the diligent refearches of Mr. Sparman we, therefore, acknowledge our obligations for the following particulars.

The natural inftinctive dread, and ftrong perception in animals in general, of a Lion near at hand, is thus related: We could plainly perceive by our animals (viz. horfes and oxen) when the Lions, whether they roared or not, were reconnoitring us at a fmall diftance; the hounds then did not dare to bark in the leaft, but crept quite clofe to the Hottentots, and our horfes and oxen fighed deeply, frequently hanging back, and pulling flowly, with all their might, at the ftraps with which they were tied up to the waggon; they alfo laid themfelves down and flood up alternately, appearing as if they did not know what to do with themfelves; and, indeed, I may fay, juft as if they were in the agonies of death. One of our oxen, on this and every fimilar occafion, appeared particularly difquieted and reftlefs, and befides, made fion, appeared particularly difquieted and reftlefs, and befides, made a remarkable noife inwardly; the fame was the cafe with the ftallion in

his particular way. At night, between 10 and 11 o'clock, we heard the roaring of a Lion, and, though it roared only twice, the animals were reftlefs the whole night through. The bounty of Providence claims our admiration in this inftance, which having fent fuch a tyrant as the Lion among the brute creation, has likewife taught them to difcern and diftinguifh it at a diftance with trembling and horror.

The following trait is curious, and, we believe, new: That the prey of the Lion, if of the brute creation, is immediately difpatched, but if of the human fpecies, although provoked, he contents himfelf with merely wounding, or, at leaft, waits fome time before he gives the fatal blow. A farmer, who had the misfortune to fee a Lion feize two of his oxen which he had juft taken out of his waggon, told me, they inftantly fell dead, though, upon examining, it appeared, their backs only had been broken: on the other hand, the converfation ran every where in this part of the country upon one Bota, a farmer and captain of militia, who had lain fome time under a Lion, and received feveral bruifes from the beaft, and was a good deal bitten in one arm; but, upon the whole, had, in a manner, his life given him by this noble animal.

The ftrength of the Lion is unqueftionably very great, yet, to ftrength he is fometimes obliged to add cunning;—to attack the Buffalo, he fteals upon his retreat, and, by furprife, feizes the animal by the noftril with both paws, which he continues to prefs clofe together, till at length, the creature is wearied, ftrangled, and dies. In running they (two which we chafed) had a kind of a fidling pace, like that of a dog, accompanied now and then with a flight bound, their necks fomewhat raifed, they looked afkant at us: they appeared to be confiderably higher and longer than our horfes, which were of the fize of common galloways.

THE ROYAL TIGER.

Drawn from Life Engraving Published by The Fenton Junr (under his Survey)   N.º D.—the Terrion Terridan Court Road Nov.ʳ 1795.

The R O Y A L   T I G E R.

T H E observation of the inferiority of works of art, compared to the productions of Nature, is particularly applicable to the skin of the Tiger, the beauty and splendour of which surpasses that of most other quadrupeds; and what the Peacock is among birds, in the opinion of the Ancients, the Tiger is among beasts. The colours which adorn the skin of the Tiger are, a bright yellow chesnut on the back, legs, and about half way down the sides, which are beautifully enlivened by black perpendicular stripes; the feet and tail are also marked with the same coloured stripes, but horizontal: the belly, chest, and cheeks are white, and also marked: is classed among the Cat Tribe.

Of Tigers there are several kinds of various sizes, but which still retain the same destructive qualities, and possess the same desire for carnage. The one here shewn is the largest of the species, and is called the Royal, or Bengal Tiger; which, Mr. Buffon says, is the only one deserving the name of a Tiger. "It is a terrible animal, and more to " be dreaded than the Lion; its ferocity is comparable to nothing. Let " us judge of its power by its stature; it generally stands four or five " feet high, and is nine, ten, and even thirteen, or fourteen feet long, " exclusive of the tail."

The one from which this drawing was made, as near as we could measure, was nine feet long, and about seven or eight feet from the rump to the nose, and about four or four and a half feet high.

Tigers, though slender in their make, possess great strength of body; the report of travellers in this particular, if not corroborated by numbers, could scarcely be credited. The Buffaloes of India are very large, yet a Tiger has been known to carry one on his back with such alacrity, that his speed was scarcely impeded; the weight of an inferior animal, or a man, will nothing obstruct his flight.

The Tiger, like all of the Cat Tribe (the Lion excepted), seizes his prey by surprise; lying in ambush, they wait with silent patience for an opportunity to spring on the destined victim, which he will bound upon at the distance of twenty or more feet; and if be chance to miss, does not persevere openly in the attack, but in a cowardly manner skulks about, and seeks another opportunity for effecting that by surprise, which a dastardly temper denies to his strength; yet, when urged by necessity, the Tiger shews either great courage or despair: In combating the Lion, he defends himself valiantly, and frequently with victory. The Tiger, if undisturbed, always first sucks the blood of his prey, rooting his head even into the cavity of the body; happiness appears at the highest when he drinks in the gore of the vanquished.

The rapacity of this animal engages him sometimes in conflict with the Crocodile, who, when the Tiger approaches the water to drink (which a constant thirst compels him frequently to), readily puts up his head in hopes of a prize; then the Tiger immediately strikes his claws into the eyes of the Crocodile; this unwelcome salute is declined by sinking in the water, the Tiger still retaining his hold. Thus circumstanced, the issue is doubtful, either combatant striving not less for victory than for life; the agonizing pain of the one, and the irritated fury of the other, producing a conflict truly savage; in which death is often the portion of both.

Instances have been known of the slightest occurrence, or noise, repelling or disappointing the attack of the Tiger. The Royal Tiger here shewn was scarcely known to the Ancients, and are rare in the East Indies, which may be considered as their native soil.

THE HUNTING LEOPARD.

The HUNTING LEO'PARD.

THIS animal is one of the many which belong to the Cat tribe, and possesses the same insidious disposition, and in proportion to its bulk, the same rapacity of temper, and desire for carnage.

The neck and belly of the Hunting Leopard are yellow, or a dirty white; the back and feet, a tawney brown, beautifully marked with black spots; the head is small and round; the ears short, as though cropped, which gives a very smart and lively appearance to the creature; is about the size of a large dog; the body rather long and narrow; the legs and tail rather long, particularly the latter—inhabits the torrid climates, as in general this tribe do, where their rapacity and unsatiated appetite secure to themselves whole tracts of country, the beauties of which man is obliged to give up to these savage possessions.

The larger Leopards may be considered of an untameable disposition, which is rarely or never conquered; severity will not reform, nor will kind usage soften; and while the Lion, the Bear, &c. may be brought to familiarity, the Leopard or Panther must ever be kept at a distance. The smaller animals of this species are, however, frequently tamed, and rendered obedient to the will of its governor; we have had occasion already to notice this in the Persian Lynx, and the Hunting Leopard may be added as another instance.

In India it is tamed and trained for the chace of Antelopes, carried in a kind of small waggon, chained and hood-winked, till it approaches the herd; when first unchained, does not immediately make its attempt, but winds along the ground, stopping and concealing itself till it gets a proper advantage, then darting on the animal with surprising swiftness, overtakes them by the rapidity of its bound; but if it does not succeed in its first efforts, consisting of five or six amazing leaps, misses its prey, losing its breath, and finding itself unequal in speed, stands still, gives up the point for that time, and readily returns to its master. The height, as it sits, is about three feet.

TIGER CAT

The T I G E R C A T.

THIS animal, in the general outline, refembles the common domeftic cat; its colour is beautiful, its fhape more flender, and
fize confiderably larger, generally between two and three feet from the
nofe to the bafe of the tail—the fur of a bright cinnamon colour, is
beautifully variegated with dark ftripes and fpots, the former along the
back and flanks, the latter mark the breaft and legs; the tail, which is
rather fhort, is alfo ftreaked; the head refembles that of the common
cat, the ears are fhort. The Tiger Cat poffeffes great activity and vigi-
lance; is indefatigable in purfuit, and bold in attacking, but fearful of
man; preys on fheep, calves, and various kinds of game. Hernandez,
in his Hiftory of Mexico, reports of the cunning and craft of this
animal, that it will fometimes lay itfelf out along the branch of a tree,
as if dead; thus exciting the curiofity of the monkey, whofe approach
is quickly followed by death. This animal is a native of America,
particularly the fouthern parts; is met with in great plenty about the
Cape of Good Hope; is frequently called the Mountain Cat, and by
Mr. Buffon the Ocelot. They inhabit both the mountains and woods.

The CIVET CAT,

IS claffed by Naturalifts among the Weafels, on account of its long body, and proportionate fhort legs; and alfo, becaufe the claws are not fecured or covered with a fheath when drawn in.

The fhape of this animal is pleafant and agreeable; the colour a brownifh grey, variegated on the body and legs, with darkifh fplotches; the nofe fharp, and black at the tip; the ears rather large and round; the tail nearly as long as the body; the length of which, from the nofe to the rump, is about 28 inches, and the height about 12 inches;—is fubject to confiderable variations both in fize and colour, which in fome is cinnamon, ftriped almoft like a Zebra.

With care, this animal will live in the more temperate climates, but muft be confidered as a native of the warmer ones; viz. the Eaft-Indies, the Philippine Ifles, Madagafcar, and the Brafils; where it produces wild in great abundance; inhabits the woods, feeds upon birds and poultry; in cafes of neceffity, eats roots and herbs; poffeffes confiderable boldnefs, and is not eafily tamed, or ever familiarized.

Of this animal great gain has been made from the perfume which it produces in a glandular pouch or bag, which fprings under the anus, and bangs between the legs: this fecretion, rated as an article of commerce, has procured an attention to be given this animal, which, but for gain, it would never have experienced. Civet, as a perfume, fome years back, was in high eftimation; many animals were kept, and fed with attentive hands, in order to increafe and fecure this exfudation of nature. In the upper external part of the pouch is an orifice, which opens into the bag where the Civet is depofited (in appearance like foft pomatum); and fuch as are kept for profit, are confined in a long wooden box, little bigger than the animal, and the receptacle of the perfume is fearched, or rather fcraped, with a fpatula or fpoon twice or thrice every week, and the produce carefully preferved; which yields a good profit.

The Dutch bring the greateft quantities of this valuable perfume to market, and theirs is generally efteemed the beft, and moft free from mixtures, which are added to increafe the weight, but impoverifh the drug. It ftill continues a confiderable object of traffic in parts of India, the Levant, and the Brafils; of which latter place Mr. Buffon doubts the animal's being a native. The fumes of the Civet, when irritated, produce a ftrong odour; and any place it rubs againft ftrongly partakes of the fcent: the fkins will long retain the valuable quality.

THE PERSIAN LYNX.

Drawn from Life, Engraved & Published by Edw. Bowen, Sec'y to the Academy.

No. 7, in the Francis Tottenham Court Road, 1 May, 1787.

The PERSIAN LYNX.

THIS animal belongs to the Cat Tribe, and is clofely allied to the Ounce and Panther; and, notwithftanding Nature has beftowed a difpofition fierce and favage, they frequently are made tame and fubfervient to the amufements of the chace. For the entertainment of the Grandees they are taken young, brought up tame, and early accuftomed to the chace of both beafts and birds. They are carried to the place of diverfion on horfeback behind a rider, and when fhewn the game, run at it with great fpeed, and, jumping on its back, infidioufly creep forward, and fcratch out the eyes; thus effecting by art what their ftrength could not achieve, they never openly attack any animal, but fuch as their fuperior force can readily overcome. One fell on a hound, which it killed and tore to pieces in a moment, notwithftanding the dog defended itfelf to the utmoft. They furprife with great addrefs the larger kinds of birds, fuch as Cranes, Pelicans, Peacocks, &c.

The Perfian Lynx is called alfo the Lion's Provider, becaufe the natives report they feek out and hunt down provifion for the Lion; this is generally fuppofed to be the caufe affigned to the effect; as they inhabit the fame climate and countries as the Lion, it is moft probable they frequently attend upon and partake of his leavings: the fame report alfo informs us, that when they call the Lion, the voice refembles that of one perfon calling another.

The Turks call it Karrah-Kulak—that is, black ears—The Perfians, to the fame purpofe, name it Siyah-gufh. Mr. Buffon calls it the Caracal.

The fize of the Perfian Lynx is fomething larger than a full-grown fox, colour not quite fo red, fur not fo long, refembling more that of an hare; cheft and belly part white, has a very long vifage, and over-hanging eye-lid, which gives a very fly, infidious and fulky appearance; the ears large and thin, infide white, black on the outfide, with a radiating tuft of hair on each of the fame. Inhabits the warmer climates of Afia.

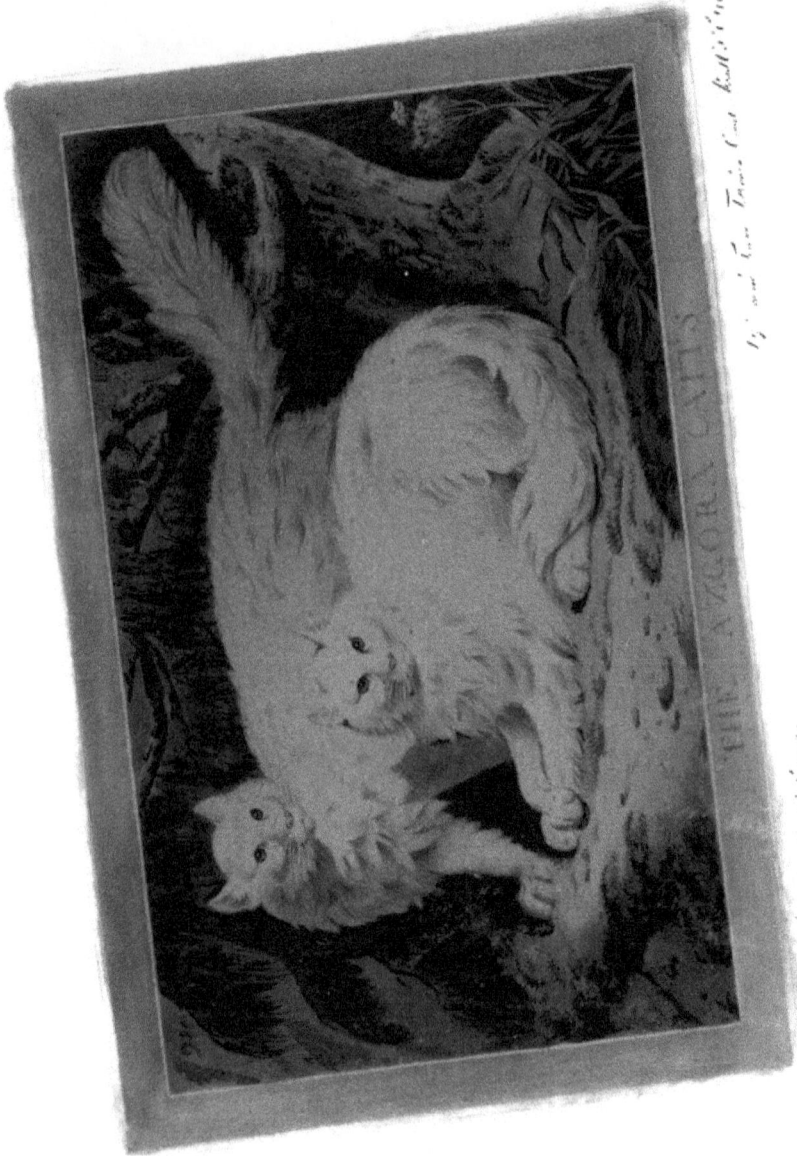

THE ANGORA CATS

The PERSIAN CAT.

IN our defcription of the Angora Goat, we had occafion to remark the influence of climate upon the hair or fur of animals; to which we fhall now add Mr. Buffon's obfervations on that phenomenon. " In general we may obferve," he fays, "that of all the climates of the habitable world, thofe of Spain and Syria conduce the moft favourably to this agreeable change of nature. Sheep, goats, dogs, cats, rabbits, &c. both in Syria and in Spain, have the fineft wool, the handfomeft and longeft hair, with colours the moft varied and agreeable; it feems as if Nature here meliorated and embellifhed the form of her animals. The wild cat of a harfh colour and a rough fkin, when domefticated, acquires a foft fkin of various colours; but in the favourable climate of Chorazin and Syria, the hair becomes longer, finer, thicker, and the colours uniformly more agreeable; the black and the red mingle to an agreeable brown, and the dark grizzle becomes a pleafant grey. In comparing the wild cat with the domeftic, we difcover no difference, but in the variation of colour. This beautiful appearance will not long furvive the change of climate; after the firft generation, they no longer attract the eye with fplendour, or invite the touch with foftnefs."

The Perfian, or Angora Cats, here reprefented, were brought from the country whence they have their name, and were in fine health and condition when this drawing was made.

The hair was three or four inches long, of a fine milky whitenefs; the eyes a lively blue; the hair on the neck longer than on the body, and the tail was bufhy; for the reft they refemble our common cats, only appearing rather larger, on account of the greater length of the fur; they poffeffed the fame habits, and were alike playful and infidious: we have every reafon to believe they were entirely deaf. They are now at Mr. Brook's Menagerie, London.

THE YELLOW–BEAR.

Drawn from the Original & Published by Cha. Catton jun.r

N.o 7 — in the Former Selebration Court Fleet Octo.r 1 1788.

The YELLOW BEAR.

OF the Bear, as well as of moſt other animals, there are ſeveral varieties, which in general are well known; the one here repreſented, was never before drawn, or deſcribed; our reſearches having not yet met with any account of the Yellow Bear.

The Yellow Bear from Carolina (as the American Bears in general are) is rather ſmaller than the European Bears; it has alſo a more pleaſant and agreeable countenance; is perfectly tame and ſociable; the colour a lively bright orange, of a reddiſh caſt. The hair thick, long, and ſilky. Its other properties are the ſame as to the ſpecies in general. This drawing was made from the animal which is now in general at London.

Of Bears in general we may obſerve, they are inhabitants of the colder parts of Europe, Aſia and America; and for the moſt part are not carnivorous, but feed on fruits, roots, berries, and vegetables; fiſh alſo form a conſiderable part of their food, in catching which they practiſe great dexterity. Of potatoes they are partially fond, and when once they diſcover them, it is with difficulty they are kept from having the greateſt ſhare. The paw of a Bear is as well calculated for rooting up the ground, as the ſnout of a hog. Are fond of peas, which they gather and beat out of the huſks on ſome hard place, and carry off the ſtraw; but tread down and deſtroy more than they conſume.

In winter they are hunted for the fleſh as well as the ſkins, which make a conſiderable article in the fur trade. Among 500 Bears killed one winter near James's River in Virginia, there occurred but two females; this being the ſeaſon for their breeding, nature dictates the moſt guarded ſecluſion, left the young ones ſhould be devoured by the males; of theſe two neither was with young.

Bears, notwithſtanding their clumſy appearance, are very nimble creatures, and will climb the higheſt trees with ſurpriſing agility, and, if wounded, will defend with great fury and reſentment to attack the aggreſſor, who, without armed aſſiſtance, has a bad chance for ſafety. In defending themſelves they ſtrike with the fore feet like a cat, and ſeldom or ever uſe their mouths, but ſeizing the aſſailant with their paws, and preſſing him againſt their breaſt, almoſt inſtantly ſqueeze him to death.

Bears, after feeding on autumn's plenty, are very nice eating; the fat is particularly mild, and more may be eaten of it, without offence to the ſtomach, than of any other fat.

Drawn from Life Engraved & Published by Tho.º Cowen f.   N.º 7 in the Terrace Tottenham Court Road   March 1.ˢᵗ 1786.

A N I M A L of the B E A R - K I N D.

THE wonderful variety of quadrupeds, and the frequent variations in each species, lead the thinking mind at once to admire the boundless productions of nature, and to confess, that man with all his acquirements of knowledge, is but very slenderly informed, and at best but superficially wise; that more knowledge and additional discoveries tend less to perfection, than to excite inquiry after a train of endless researches.

The animal we are now to describe is of a form unknown, of a species never described, with peculiarities which have never presented themselves, or else have escaped notice.

The "PETRE BEAR," (which is the name we have been directed to call it by a very able naturalist) was brought to England in August last, by Capt. Pearson, on board a ship belonging to the India Company, whose report is, that it was brought from Patna in the province of Bengal, and when it first came into his possession was very young, nearly a cub; —here information leaves us: the manner in which it was caught, or the place and circumstances attending, were not related; we must therefore rest contented with a description of its several parts, noticing such peculiarities, as have been observed since it came to England. The animal is a female, and is now to be seen at the beast shop, Holborn-hill, London.

The Petre Bear (if the name be allowed) has five claws on the fore feet, between two and three inches long, nearly of the same thickness, and not very bulky: the hair all over the body is of a rustyish black colour, very harsh and coarse, between twelve and eighteen inches long; that on the shoulder rather longer, which it can draw forwards, or lay backwards: the form of the head, and chiefly the mouth, is the principal characteristic of the animal; this we have endeavoured to ren-

der as intelligible as possible, by adding a view of the mouth when open: from the eyes to the tip of the nose, is about 6¼ inches, which tapers off like the truncated snout of swine: the front of the mouth, when shut, is flat also, like the swine: the formation of the nostrils differs from every other animal's, the natural shape being exactly as shewn in the drawing: the lips project very far (two or three inches) beyond the front teeth: the lower lip, as well as the upper one, with the nostrils, are very flexible, over which the animal has great command, placing them in any position at pleasure; has great power of suction, and will draw things into its mouth at a considerable distance: the tongue is rather small: teeth at present few: was fed, on board the ship, with boiled rice and sugar; at present on bread (about a quartern loaf per day): with some difficulty was brought to eat flesh, which it does now with a relish: is very tame; appears pleased when kindly noticed, soliciting play: expresses anger with a kind of accented growl, something like barking: its paws, when laid together, it sucks with a tremulous noise, like one when shaking with pain: the feet are short; has nails on the hind legs: drinks, or rather sucks water in considerable quantity: general appearance black, face grey, stands about 2¼ feet high, 5 feet long, has grown considerably since it has been in England.

Upon the whole, this may be considered as a very uncommon creature, as well for its formation, as the very extraordinary circumstance of so thick and coarse a fur to a native of so warm a country; the general appearance also is that of a Bear, whereas we know of none in those climes, nor any animal whose appearance can lead us to suggest to what breed it belongs, or by what mixture or chance it is likely to owe its birth. Time and further inquiry only can clear up these at present singular facts.

THE SWEEDISH — ELK.

Drawn from Nature & engraved & Published by Chst: Catton Jun.̃   N 15 in the Tower   Tottenham Court Road Oct 1. 1788.

The E L K.

T HE animal from which this drawing was made, was a native of Sweden, the dimensions and other particulars of which we shall first enumerate.

The length of the head was two feet, the body three feet, and neck one; the total height was six feet, of which the legs were three; the ears were about twelve inches long; the general colour a blackish brown brindled, some hairs being brown, some white, and others partaking of both colours; the lightest colour was on the neck, and upper parts of the body; the knees of the fore legs, and inner parts of the hind legs white, which extended but little on the brisket; the hair universally very coarse and thick; about two inches and a half long; considerably longer under the lower jaw, and along the upper part of the neck, resembling an upright mane, which was longest and fullest over the shoulders; the tail not more than two or three inches, was nearly lost among the hair of the body; the head long; the upper jaw about the nostrils very full and chumpy. We have been the more exact in the particulars of this animal, few having been seen in England; we have knowledge of two only, one belonging to his Grace the Duke of Marlborough; the other now exhibited at Mr. Parkinson's Museum, London, from which this drawing was taken.

Of the general history of the Swedish Elk, the following particulars have come to hand. In walking, the Elk carries its head nearly horizontal, and when passing the thick parts of the forest, the head is so disposed that the horns lay close upon the shoulders; thus the inconvenience of such wide extended antlers is much avoided. The horns of this animal are reported to grow to very surprising dimensions; the

fossil horns frequently found in Ireland are supposed to belong to the Elk, or an animal of that species: some have been found extending fourteen feet; those to the animal here represented, were distant about four feet at the top; each horn having five tips, leads to suppose the animal was about six years old, and certainly not yet come to its full growth. The climate of Sweden in general, but of the northern parts in particular, being severely cold, nature has not only provided a thick warm fur, but has so formed the under part of the hoof, with a very sharp edge, that this animal with perfect safety can pass over the smoothest ice. The race is not always to the swift, nor the battle to the strong, else this animal would be secure from harm; the bear and the wolf hunt it with success, the shape of their feet, and the lightness of their weight, enabling them to travel the frozen snow with safety; while the flight of the intimidated Elk is impeded by frequently falling through the frozen crust, which not only stops its progress, but also wounds its heels; thus suffering from pain which is increased by repeated disasters, it happens ere long the enemy seizes it by the throat, and the largest European animal is conquered by one, not a quarter of its size, or weight. This animal, when defending itself, uses not only its horns, but rising on the hind legs endeavours to plunge the fore ones into the body of its adversary.

It feeds upon herbage, prefers the aspen tree, resides on uninhabited islands, in large rivers, and on the banks of great lakes. The report of its speed is, that it will travel 300 miles a day; the flesh rather brown and coarse, but good tasted.

The A N T E L O P E

FORMS a diſtinct claſs of animals, partaking many of the charac-
ters of the goat and the deer, yet diſtinct from either.

The Antelope, or Gazelle, has hollow, truncated and permanent horns,
which is not ſo in the deer; declines the graſſy paſture, and browſes on
ſhrubs, which imparts to the fleſh a delicate and agreeable flavour; on
the other hand, the ſize and delicate form agree with the deer; the
colour and nature of the hair alſo is the ſame, but the form of the horns
differs very much, being of one ſtem or ſtalk only (without any branch-
ings out), though of various and very different inflexions, which
mark the ſeveral ſpecies of this tribe. The horns are annulated or
girt round with rings at various intervals, at the ſame time they are
longitudinally depreſſed or flattened from top to bottom; theſe particulars are common to all of the numerous race of Antelopes; of which
Mr. Buffon has enumerated twelve; Mr. Pennant doubled that number,
and perhaps there may yet be many undeſcribed varieties, if variations
ſo flight, and marks ſo ſmall, conſtitute a new kind. The influence of
ſoil, country and climate, has moſt likely produced this great variety,
and where it may end is too uncertain to ſay. In general, they inhabit
the hotter climates of Aſia and Africa (three or four kinds only excepted), and go in companies of ſix or eight, or elſe herd together in vaſt
quantities.

Mr. Sparman mentions a lucky eſcape he and his travelling equipage
experienced, in being about a quarter of a mile on one ſide a herd, not
leſs than ten thouſand, which took their courſe over the plain where he
lay encamped, and, but for this fortunate ſpace, he and his compa-
nions would have been trodden to death.

The eyes of this animal are the moſt beautiful and meek which
Nature has formed. The eye of the Antelope is a metaphor in general
uſe among the Eaſtern poets; and the gallantry of a lover, in thoſe
countries, can go no higher, than in comparing the eyes of his miſtreſs
to thoſe of the Antelope.

The form alſo is very elegant, and, with their ſwiftneſs, is noticed
by the Sacred Writers. Their activity is wonderful, and their ſpeed
exceſſive; the fleeteſt dogs are left far, very far behind; and when
hunted, the aid of the hawk is neceſſary, which, by faſtening on the
neck and check, either mortally wound, or ſo much impede their
flight, that an opportunity is given for the horſemen and dogs to come
up and ſecure the game. The Lynx or Panther is often employed in
this chace, for the amuſement of the Great; theſe animals ſucceed more
by craft than ſpeed; creeping ſlyly forward in a winding courſe, at an
unwary moment they ſpring upon the thoughtleſs animal, mortally
wound, and ſuck the blood.

Antelopes are very timid; ſome dwell upon the plains, others, and
moſt generally, among the hilly countries; at the leaſt or moſt diſtant
alarm betake themſelves to flight, and ſeek for ſafety in dangerous and
inacceſſible parts of the rocks; where, it is reported, the Antelope will
ſtand upon a pinnacle or point, no broader than the ſpace occupied by
the four feet drawn cloſe together.

The one here delineated is the common or brown Antelope, the
horns of which are about eighteen inches long, and fifteen diſtant at
the tips; ſtands about three and a half feet high; colour a bright brown;
cheſt, belly, &c. white.

THE WHITE-FOOTED ANTELOPE.

The WHITE-FOOTED ANTELOPE.

THE great variety of Antelopes, and the material difference be-tween this and the one before given, induced us to prefent our readers with this and this drawing of the White-Footed Antelope. In our for-mer account we remarked the almoft innumerable varieties of this tribe, with marks effentially different, yet partaking of the fame gene-ral character and properties.

The Antelope now before us, is marked by four white feet; the general appearance alfo varies confiderably; the colour is a dark or blackifh brown; on the cheeks are two white fpots, and on the neck there is a tuft of black hair.

Its ftrength and activity were very great; much pains and labour were beftowed before it could be brought to be the leaft fubmiffive or familiar; its height was equal to that of a common galloway, or fmall horfe; the hoofs were long, and divided far up.

Having before prefented our readers with the leading features and principal characteriftics of the Gazell or Antelope tribe, we beg to refer to a comparifon of the two drawings, for the more minute variations of thefe two animals of the fame fpecies.

THE MUSK DEER.

Drawn from Nature & engraved. Publish'd by Cha.º Catton, Jun.º

N.º 5 in the Tower Fartan Grove Road.

The MUSK-DEER.

OF the Musk Deer much doubt and various opinions have been held, by travellers and naturalists, in which class of quadrupeds to place it; some contend for the Horse, some the Ox, and other the Deer; the latter appears the best decision, notwithstanding the animal wants those striking characteristics the horns; suffice it for us, the animal is known by the name of the Musk-Deer. The figure here drawn, will better shew its form, than any description.

This animal is particularly known by producing that celebrated drug the musk, which heretofore was in high esteem as a perfume, and at this time is considered as a very powerful and valuable medicine in nervous and other cases. In the male animal only, this exudation gathers in a small bag or cushion, between the legs, near the groin, in substance like coagulated blood, or, as some describe it, a brown fatty matter; indeed certain information is wanting of this particular. Mr. Gmelin reports, that when the bag by being over-full becomes troublesome, the animal expresses part by rubbing against a tree, or other convenient place; the matter dropping in small parcels on the ground is secured, and is esteemed by much the best musk.

Others again report, that it comes to market only in the repository formed by nature, which is not bigger than a small egg, and that the weight of four is not more than an ounce. Considering its smallness, the animals must be in great abundance to supply the quantities used in Europe only; in the East its reputation being much greater, its use must be more general. The male animal only producing this treasure, must be hunted and killed for the bag. This drug, like every other article of value, is very liable to adulterations; small pieces of lead are thrust into the bag to increase the weight, and foreign matters often mixed with it, to augment the quantity.

To the mouth of the male animal there grow two teeth or tusks from the upper jaw, rather bent; these are very white; to the female these are wanting; the latter also is much smaller than the male, The flesh is strongly impregnated with the perfume of musk, but is nevertheless eaten by the Tartars.

They inhabit various parts of the East Indies, China, Tartary, and Siberia, from which latter place the one here drawn was brought. The colour, a reddish brown, marked on each side the throat with a stripe of white, and several splotches of the same on the flanks; the attitude of the hind legs was also very aukward, as expressed in the drawing. Naturally dwell among the mountains, and feed on herbage, and the young buds of trees, particularly pines: prefer solitude, and avoid mankind: if pursued, flee to the highest summits, inaccessible to men or dogs.

The size of the one from which this drawing was taken, was, from the nose to the insertion of the tail, two feet four inches; from the ground to the shoulder, twenty inches.

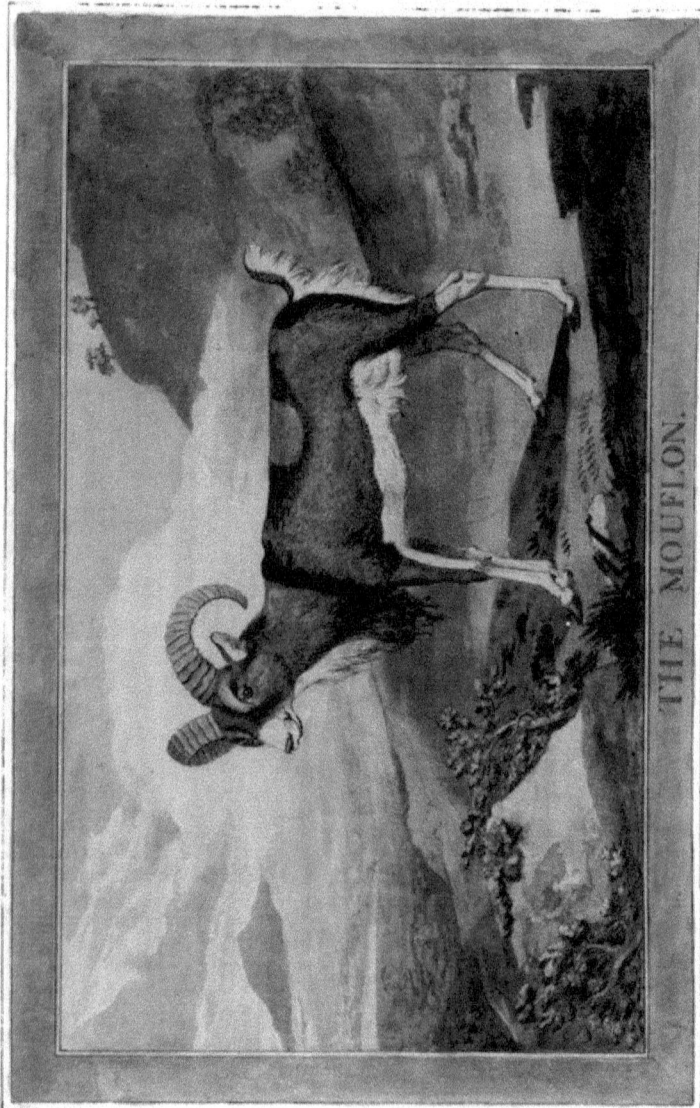

THE MOUFLON.

Drawn from Nature Engraved & Publish'd by Chr. Becton Jr.

N.º Van Esmore Tottenham Court Road Novr. 1 1795.

The MOUFLON.

NATURE with a wise and provident care gave to every animal originally either force to repel, speed to escape, or cunning to evade its more formidable enemies, as a means of preserving the continuance of her various productions. Of useful animals the weak were first reduced to the services of man, smaller animals more readily adopting the influence of education. Thus the Sheep and Goat were first brought to usefulness, before the robust Ox, or vigorous Horse; this may be considered as one principal cause of the endless variety of such as have been long and particularly attended to by man; and the power of continued education, joined to the influence of climate and soil, make it difficult to say, which are the true characteristic marks, or which the original of many species.

The Mouflon is considered as the Sheep in a state of nature, by Mr. Buffon, with a temper not broken by servitude, a constitution not softened by inactivity and luxury. With a vigorous mind it defends itself against the attacks of larger animals, and, aided by a robust body, frequently overcomes formidable enemies.

The abode of the Mouflon is in rocky countries, where they bound from rock to rock, or climb the apparent inaccessible precipices with that address and ease, which characterize the Goat and Deer tribes, and which sets pursuit at defiance. The horns of the Mouflon are very broad at the base, are firmly fixed upon the skull, inclining backwards with a considerable curvature, the distance increasing to the extremities; the horns, of a light brown or yellow, are girt with many annuli or rings; in the male they frequently grow to very large dimensions, and weigh sometimes thirty pounds; and, when broken off, as in defence, or by other accident, often serve as a nest, or retreat, for various small quadrupeds, such as young foxes, &c.

There is a beautiful form in this animal, which approaches very close to the Deer; indeed by some it is esteemed the same as the old or original Welch Deer; however, the shape of the head, and the truncated horns, which are never shed, rather mark it of the Sheep tribe. The major part of the body is of a fine brown coloured hair, which on the lower part of the neck and chest grows to a considerable length; the belly, legs, and outer part of the haunches are white.

Inhabits the warmer parts of Europe, such as Greece, the Isle of Cyprus, Sardinia, and Corsica; they also are found in great numbers in the southern and mountainous parts of Siberia, a climate rather cold than temperate; grows to the size of a young stag. The measurement of the one here drawn was three feet from the nose to the rump, and two feet and a half from the shoulder downward.

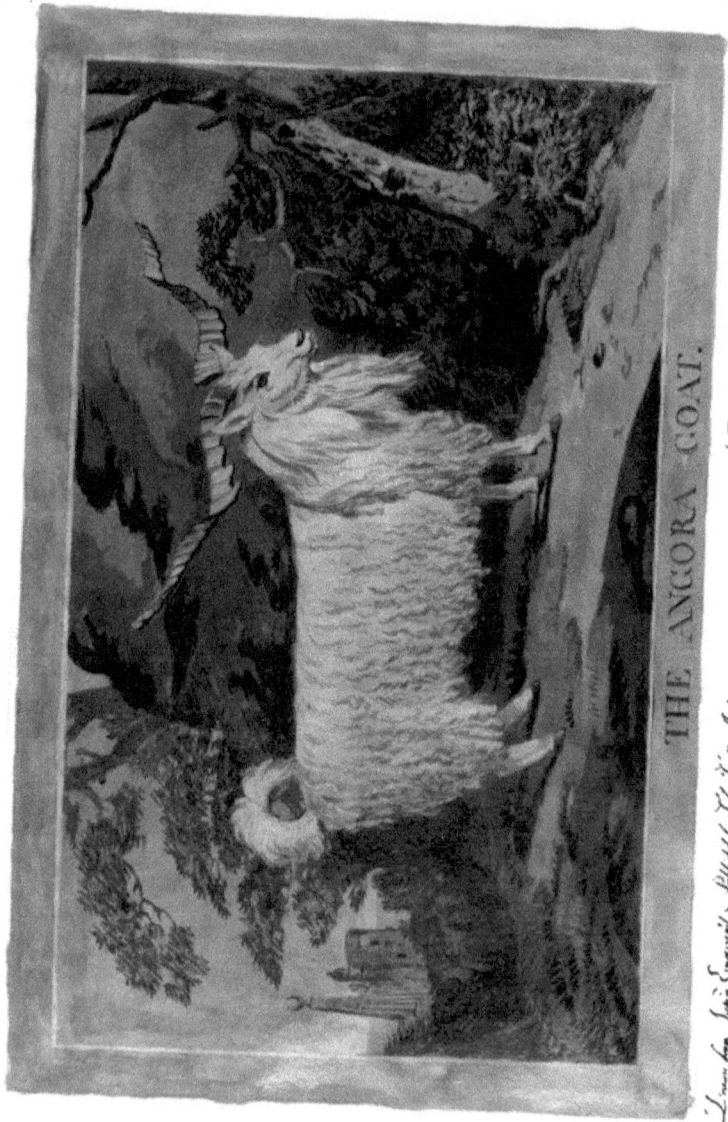

THE ANGORA GOAT.

The ANGORA GOAT.

THIS animal affords the moſt ſtriking example of the influence of climate upon the clothing or fur of Goats and Sheep. When transferred from a cold or temperate country to a hotter one, the fine wool of the ſheep degenerates to a coarſe hair ; and a few miles of country ſurrounding Angora in Aſiatic Turkey, produces this very extraordinary variation from the common appearance of Goats. The climate of theſe countries, and Syria in particular, impart to all the quadrupeds, a ſplendour and fineneſs of fur, which is not to be equalled in any other part of the world.

The horns of the Natolian or Angora Goat are of great length in the male, and lie in an horizontal poſition, twiſted like a cork-ſcrew ; to the female the horns are very different, the form nearly a complete circle round the ear.

The hair of the Angora Goat exceeds in beauty that of every other animal ; the ſilvery locks formed by Nature's careleſs, though graceful hand, flow in ringlets of about nine inches long, with a ſilky fineneſs, and reſplendent whiteneſs.

The hair of this Goat, for its beauty and fineneſs, is bought by all nations, and is the baſis of our beſt camblet ſtuffs ; it is alſo wrought into the uſeful article called mohair. Mr. Pennant reports, the hair is imported in the form of thread, as the Turks will not allow it to be exported raw, as the ſpinning gives employ to a multitude of poor. Much pains is taken by the owner of the flocks, in keeping the animals clean, and often combing them. Of late, ladies muffs in England have been frequently made of the hair of this animal, the delicacy and graceful form of which produces a pleaſing and agreeable effect.

The one this drawing was made from was conſiderably larger than our Goats ; the meaſurements were three feet ſix inches from the noſe to the rump, and two feet two inches from the ſhoulder downward.

The H Y AE N A .

I S in length about four feet, in form between the wolf and hog; the head resembling the wolf's, but rather shorter, and black from the nose to just above the eyes; the colour a greyish brown, marked down the body and legs with darker stripes, inclining to black; stands remarkably high before, and low behind, being in front about three feet and a half, and behind about two feet and a half high; is very lank or thin in the body, with a ridge of bristles like a hog, all along the spine, and a brush tail.

The Hyæna is a remarkably unsociable and solitary animal, dwelling in the holes and chasms of rocks, or in dens which it has formed in the earth; it is found in high, mountainous countries, in the mountains of Caucasus, the Atlantic chain, parts of Syria, Persia, and Barbary, the most dreary and sterile parts of the Torrid Zone, of which, says Mr. Buffon, it is a native.

This animal possesses great strength of body, and fierceness of courage; it will resist the attacks of the Lion—(Kæmpfer relates that he saw one which had put two Lions to flight)—he never declines a combat with the fiercest of the forest, and seldom fails to conquer—possessing great cruelty and fierceness, he is generally reckoned untamable. Mr. Buffon mentions to have seen one, and Mr. Pennant another in a domestic state; the latter Gentleman thinks, if taken very young, they may be reclaimed by good usage, but observes, they are commonly kept in a state of ill humour by the provocations of their master.

Hunger seems ever to torment them with an insatiable voracity; they greedily devour whatever comes in their way, and will even root up the contents of the grave, and with a keen appetite consume the putrid corps.

The ancients had very strange notions of this animal, believing that it changed sex every year; this opinion, no doubt, originated from the transverse orifice which is between the tail and the anus; they also believed that a stone was found in the eye which imparted the power of prophecy. Mr. Shaw, in his travels into Barbary, observes, the Arabs always bury the head of this animal, whenever they kill one, lest it should be applied to purposes of magic; it was also believed, the Hyæna had the property of imitating the human voice, and thereby seduce unwary travellers within its power; this the experience of modern times does not confirm, yet a gentleman told us, he once heard one make a noise resembling the laughing of a man, when the keeper had just given him some provision, which he made as if going to take away—this was, probably, no more than a note of displeasure, which every animal possesses, such as the growl of a cat with a mouse when any one approaches.—To conclude, no words can convey an adequate idea of the voracious and fierce aspect of the Hyæna, or of his uncouth, ill-formed shape.

The W O L F.

THIS animal bears a strong resemblance to the dog; it has a long head, a sharp nose, pointed and erect ears, a bushy tail, hanging more between his legs than a dog's generally does, long legs, and large sharp teeth.

This animal abounds in every quarter of the earth, it inhabits large extensive forests, frequently making incursions on the campaign country in search of prey; its food is such animals as its superior strength can vanquish; lambs, calves, and sheep, which he carries off with ease and agility; the muscles of his neck are so strong, that in running off with a large sheep, it shall not trail on the ground. This animal is remarked for great cruelty; whenever he gets among a flock, he destroys with a wanton pleasure all that come in his way, before he secures his particular prey. It is reported, if a Wolf once tastes human flesh, he ever after prefers it, and will seek it with great avidity and resolution, and sometimes attack the shepherd in preference to the flock. Troops of this destructive animal have been known to attend the march of an army, and with eager appetite devour the slain in battle, or such as were superficially interred:—in general, this animal pursues his prey alone, but when any great business is to be achieved, such as to attack an ox, deer, or other large animal, they will assemble in numbers, and commit their depredations in concert; but when the business is finished,

the assembly breaks up, and each retires to his hiding place. The scent of this animal is remarkably good, as he will wind his game at many miles distance. To hunt the wolf is esteemed good sport; dogs of great courage and speed are requisite, as they sustain a long chase. When pressed with hunger, which he long sustains, if well supplied with water, this animal exercises great courage and cunning: The destructive properties of this animal make him every where obnoxious; two or three will cause a long and constant alarm to a whole country; enmity therefore between him and man is perpetual. England abounded with this scourge of rural life, till king Edgar respited the punishment of a criminal for a certain number of wolves' tongues, and changed the tax of gold and silver paid by the Welch into an annual tribute of three-hundred wolves' heads. King Edward I. entirely to extirpate them, appointed a superintendant to assist in their destruction; which was not effected in Scotland till about the middle of the last century. The sagacity of the wolf is so active, that, Scheffer reports, he will not attempt an animal, if a cord, or halter is round its neck, suspecting a trap. The wolf is about the size of a large dog, or three feet and a half long, and two feet and a half high; colour a light brown, with a mixture of black and grey; has a penetrating eye of a grey colour, which adds a wild appearance to the ferocity of his countenance.

THE OTAHEITE-DOG

The OTAHEITE DOG.

THE modern discoveries in the South Seas, which have been so accurately reported by that celebrated navigator, Captain Cook, have, with a new track for speculation, opened a new field for observation; and the productions of these isles have been fought after with as much curiosity and impatience by some, as a buried vase from Herculaneum, or the remains of the Capitol by others. Under this persuasion of the pleasure of novelty, we present our readers with a portrait of an Otaheitean Dog; which, as an animal of infinite importance, as an article of luxury with these people, may deserve some notice from us.

The Otaheite Dog is about the size of our large spaniels, and nearly resembles them in appearance; the head is rather longer and deeper, or flatter perpendicularly; the ears are erect like the wolf's; the limbs appear rather larger; the colour, for the most part, white, with lively brown spots or blotches.

The estimation of things in general depend much upon their abundance or scarcity; and with an Otaheitean, whose quadrupeds are but two, it will not excite much surprise, that these are attended to with some anxiety, particularly so, when the pleasure of the palate is concerned: this is an influence to which the most savage nations pay respect.

To give some idea of the importance of a dog in the South Seas, we shall present the report of Captain Cook on this particular, as given in his first voyage: "We all agreed that a South Sea Dog was little inferior to an English lamb; this excellence is probably owing to their "being kept up, and fed wholly upon vegetables." Thus much as evidence of their delicacy; the manner of cooking this dainty shall close our account. Of one presented to the Captain, we read, "the "dog was killed by holding the hands close over his mouth and nose; "an operation which continued about a quarter of an hour: while this "was doing, a hole was made in the ground about a foot deep, in which "a fire was kindled, and some small stones placed in layers, alternately "with the wood, to heat; the dog was then singed by holding him over "the fire, and, by scraping him with a shell, the hair taken off as clean "as if he had been scalded in hot water; he was then cut up with the "same instrument, and his entrails, being taken out, were sent to the "sea, where, being carefully washed, they were put into cocoa-nut "shells, with what blood had come from the body. When the hole "was sufficiently heated, the fire was taken out, and some of the stones, "which were not so hot as to discolour any thing that they touched, "being placed at the bottom, were covered with green leaves; the dog, "with the entrails, was then placed upon the leaves, and other leaves "being placed upon them, the whole was covered with the rest of the "hot stones, and the mouth of the hole close stopped with mould; in "somewhat less than four hours it was again opened, and the dog was "taken out excellently baked, and we all agreed that he made a very "good dish.

"The dogs which are here bred to be eaten taste no animal food, "but are kept wholly upon bread-fruit, cocoa-nuts, yams, and other "vegetables of the like kind."

The G R E A T B A B O O N.

BEFORE we enter upon the defcription of this creature, it may not be amifs juft to premife, that Naturalifts have divided the Monkey tribe into three claffes; viz. Apes, Baboons, and Monkeys.

Baboons are characterifed by having fhort tails, canine teeth, and the lower part of the face prominent or truncated, like the fnout of fwine: In general, they, with the reft of the Monkey kind, go upon all fours; though fometimes, and the Ape more frequently than the Baboon, or Monkey, go erect: the pofition defigned by Nature was certainly prone, and when in their wild ftate, and uneducated, this is evidently their manner of walking; elfe would the fore feet or paws have more foftnefs of fkin, and lefs callofity than they really have; in this particular the fore feet differing nothing from the hinder ones, which muft neceffarily be conftantly trodden on: the fingers of the feet are armed with long fharp claws or nails.

Of Baboons it may, in general, be faid, they are fierce and ferocious, of great bodily ftrength, perverfe habits, and very libidinous: extraordinary ftories are related of their partiality to, and anxiety for, the females of the human race.

The one here reprefented is called the Great Baboon, and was in height about 5 feet; from the head to the rump was about 2 feet 9 inches; the length of the arms about 1 foot 9 inches; general colour a greenifh black, mixed with brown; acrofs the hams a patch of purple; a bright vermilion ftripe up the nofe to the eyes; the cheeks a dark violet blue, with horizontal ftripes of white; the eyes fmall, and approach very clofe to each other; long hair on the head and fhoulders in confiderable quantity, a little brighter than the reft of the body. The inner part of the cheeks, that is, the fpace between the cheek and the teeth, is very capacious, and ferves this fpecies of animal as a pouch, in which to ftow provifions, &c. Feeds on roots, fruits and herbs, naturally is not carnivorous, eagerly fond of fpirituous liquors and wine—inhabits the woods of the hotter parts of Africa.

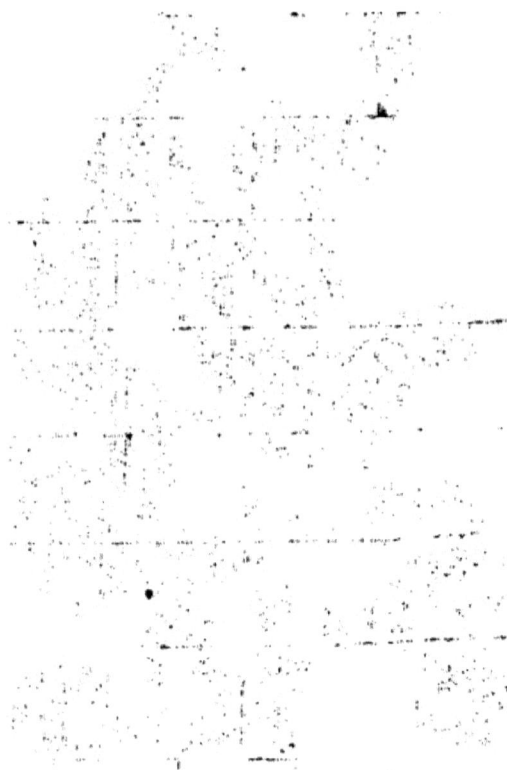

The LION MONKEY.

T HE Monkey tribe is divided into three claſſes—Apes, Baboons, and Monkeys.

Monkeys are characteriſed by having a long tail; and are again ſub-divided, by Mr. Buffon, into two claſſes; diſtinguiſhed by the uſe they make of their tail in many of the friſky manœuvres and tricks this ſpecies of animal is ſo famous for. To the one the tale is called prehenſile, and ſerves almoſt the purpoſe of another hand; for this they readily twiſt round the branch of a tree, and by it ſuſpend themſelves hanging in the air, head downwards; it alſo ſecures them in their ſeat, while the feet are otherwiſe employed. The other kind do not enjoy this uſeful property of the tail; which, Mr. Buffon has obſerved, belongs to none of the Monkey tribe of the old Continent: on the other hand, thoſe of the old world poſſeſs a cavity or pouch on each ſide the jaw, which ſerves as a ſtore-room for proviſion, and which thoſe of

America do not enjoy. Thus the diminiſhed activity in the uſefulneſs of the tail is balanced, by an opportunity of laying in ſtore that, which elſe, might not at all times be ſo readily acquired.

The Monkey here repreſented is called the Lion or Silken Monkey; the colour and appearance of the hair about the ſhoulders reſembling that of the Lion; the hair over the whole of the body is long and very fine, with a moſt beautiful and ſilky appearance; the tail is long. This animal, when ſitting, did not exceed 10 inches in height.

It is the practice of the Lion Monkey to take up his abode in a large melon, or gourd, which having previouſly excavated, and lined with ſoft cotton, forms a comfortable habitation.

It is a Native of South America, particularly Guiana, and the Braſils; is rather delicate, but gentle and frolickſome.

THE CHILD of the SUN

Drawn from Life Engraved & Published by Chas Catton Jun.r N.D on the Terrace Tottenham Court Road Dec.r 1794

The CHILD OF THE SUN.

THIS animal is hitherto a non-descript, belonging to the class of Baboons.

The only one we have knowledge of (the one from which this drawing was taken) was exhibited in England about four years back, and was reported to have been brought from South America. The head, in proportion to the other parts, was remarkably large; the contour or shape of the face also rather singular, for one of the Monkey Tribe; the skin of the face smooth, of a fallow complexion. The height of this animal, when erect, was five feet, and to its great stature was joined great strength of body; the hair from the shoulders all over the body was very long, but not coarse; the colour a light speckled grey, resembling a Guinea fowl.

The cunning and subtlety of the Monkey was very apparent, and in all its actions it closely imitated the human: the haims or buttocks were bare, and of a bright vermilion colour. The skin of this animal, when dead, was deposited in the Leverean Museum, where it is now to be seen.

Having already noticed the characteristics and leading properties of Baboons, we shall here add some further general account of them.

"The Baboon is a gregarious animal, herds together in great numbers, and mutually unite their strength to repel danger, and procure subsistence. The œconomy of Baboons, in general, is well regulated, and those of the Cape of Good Hope, we are informed, observe a sort of natural discipline, and go about whatever they undertake with surprising skill and regularity; not being carnivorous, an herd of hundreds consume great quantities of fruits, &c. Grapes, apples, and garden fruits in general they are particularly fond of, and when they set about robbing an orchard or vineyard, centinels are always placed to give early notice of the approach of danger; these necessary precautions taken, the plan of operation is as follows: part enter the inclosure, pluck the fruit, and chuck it to their nearest fellow without the fence; a regular line of communication being formed from the scene of operation to the place of retreat, the plunder is pitched from one to another all along the line, till it is safely deposited at head quarters, which usually is in some mountain. During these manœuvres the centinels keep a close look out, and if danger approaches, a loud cry is the signal for retreat; this is done in a very quick, but not improvident manner, as each one loads himself in the mouth, hands, and under the arms: if closely pursued, the latter parcel is first dropped, next that in the hands, and last of all, if very much pressed, that in the mouth. They commit their depredations with such boldness and address, that the natives, to protect their property, are subject to frequent watchings, and nevertheless suffer great damage." *Kolben's Cape of Good Hope.*

THE MAUCAUCO.

The M A U C A U C O.

THIS animal claffes in the rear of the Monkey tribe, and ferves as one, to connect the gradation from that fpecies to the complete quadruped, and belongs particularly to the clafs called Maki. The hands are ufed the fame in this fpecies as in the complete Monkey, ferving for every purpofe of feeding, climbing, and playing.

The nofe of the Maucauco is long and flender, black on the tip; the eyes very large and fine, furrounded with a circle of black hair; the ears large and upright; the arms, feet, and toes, or claws, like the Monkey's; the tail beautifully marked with alternate black and white rings, and is confiderably longer than the body, which is flender, of a pale brown, or afh colour, fomewhat darker along the fpine; neck and belly white. The hair is beautifully fine and foft, and ftands erect, nearly as the pile of velvet.

The Maucauco is a native of Madagafcar, and the neighbouring ifles; is very good-natured and frolickfome; poffeffing all the motion and alacrity of the Monkey, without its malice or mifchief, and is very cleanly; has a weak cry, is eafily tamed; in a wild ftate go in troops of thirty or forty.

THE ERMINE.

The ERMINE.

THIS little animal is pretty well known on account of the high estimation in which the skin is held. The fur of Ermine is an article of considerable commerce in the more northern countries; its delicate whiteness and closeness being surpassed by none.

The Ermine is about nine inches long, exclusive of the tail, which is between five and six inches more; the tail is black, and in the male part of the forehead is dark brown.

The animal called in England the Stoat is the same as in more northerly climates is called the Ermine. It is a curious phenomenon in Nature, that furs of animals in general, in those countries which have a long and severe winter, at that season change their colour; thus, in Siberia the Fox, &c. are white, and among others the Stoat or Ermine also undergoes the same change; from a skin of brown and dark yellow, thinly covered, is produced that thick sett, pure white fur which is prized by all, and bought at a very high price by many nations. The process or manner how Nature accomplishes this metamorphosis, is not so easily traced:

we well know her maternal care does accomplish it, as a means of protecting the life of her offspring by a sameness of colour with the snow, which for months covers the ground where this little animal is placed to endure the rigours of severe frost; and, although their escape is thus facilitated, the wily arts of man succeed in the destruction of thousands, by means of traps of various kinds baited with flesh. The hunters of Norway shoot them with blunt arrows.

In very severe seasons the Stoat in England (and very frequently in the further parts of Scotland) is known to change white; but their skins are of little value, the severity of our winters not being sufficient to effect so complete a change in the colour and substance of the fur, in which its greatest merit consists.

An Ermine brought to England quite white in the month of May, lost all its splendour in about thirty days (beginning at the head), which did never return; its food is rabbits, birds, mice, &c. is very quick of motion, and has a fœtid smell, as all of the Weasel Tribe have.

ANIMAL of the WEASEL-KIND.

OF the limited knowledge of man, and the unlimited bounds of the animal creation, another inftance occurs in the fubject we are now to defcribe. This animal muft be confidered as a non-defcript; the regularity and proportion of the features are fuch, that it appears a perfect animal, that is not a variation produced by chance, nor the offfpring of a mifcellaneous copulation.

The body is twelve inches long, and with the legs and head is black; on the back are four longitudinal, broad, waved ftripes of white; in the front of the forehead is a fmall triangular fpot of the fame; the ears are fhort and round, and white on the infide, which is continued a little down the face; the tail is 10 inches long, very full and bufhy, the hair foft and fine; the tail in general is down, but erect, as fhewn in the drawing, when pleafed or frightened; has five claws on the fore legs, and four on the hind ones; teeth are very fmall and fine; fleeps in the day time, at prefent in a lady's fkin muff; when awake in continual motion; is very agile and frolickfome; is very tame and docile; anfwers to the name of "Jack," and readily comes to any one when called: legs rather fhort.

Was brought from Bengal laft fummer, by Capt. Gell, of one of the King's fhips; is now in the poffeffion of Sir Jofeph Banks, by whofe permiffion this drawing was made.

Upon the whole, the tail of the animal, when erect, is like that of the Squirrel; but, from the fhape of the body, we think it more properly claffes with the Weafel tribe: indeed, the name handed to us was "THE STRIPED POLE CAT."

THE BROWN - COTI.

The B R O W N C O T I

IS an animal of the Weafel tribe. There are feveral varieties of the Coti; the one here reprefented is the fame as Mr. Buffon names, *Le Coati noirâtre.*

The Brown Coti has a longifh head, the fnout confiderably elongated, but not fo much as in one of this fpecies of animals. The ears are fhort, tip of the nofe very flexible, and of a purple colour; a light ftripe from the nofe to the back of the forehead; cheeks almoft white, with a patch of white above the eyes; feet fhort and black; eneral colour brown, mixed with nearly black hairs; the fur coarfe

and long; length of the body about three feet from the nofe to the infertion of the tail, which is about two feet fix inches more; the neck and body long.

. The Cotis are natives of South America, particularly Brazil and Guiana; feed on fruits, eggs, and poultry. The one from which this drawing was made, was very greedy. Run up trees, or any perpendicular place, very nimbly; eat like a dog, holding the food between the fore paws; but drink by fuction like a pig; are eafily made tame, and are much inclined to fleep in the day-time.

THE BADGER.

Drawn from Nature & painted by W. Hastings. — W. Peacock, fecit.

N.Z. on the Terrace, Tottenham Court Road. June 1 1799.

The B A D G E R.

THE Badger, though a native of England, is an animal not very well known; the fhynefs of his temper, and ftrong propenfity to fleep, which foftens the call of appetite, makes his appearance in fearch of prey lefs neceffary and frequent; the night alfo is the feafon for his fearch; when, whether vegetable productions are fatisfactory, is not certain. The depredations in rabbit-warrens, and on young lambs, are frequently laid to the charge of the Badger; on the other hand, the Badger has little or no fpeed, and being much inclined to fleep, will certainly grow fat on lefs nourifhment than more active quadrupeds.

The Badger digs a habitation in the earth with confiderable dexterity; the fore paws being armed with long and ftrong nails, which work with great expedition: the paffage is of a winding form, leading to feveral apartments, but only one entrance; a bed of foft hay and grafs is provided to induce fleep, and fecure a comfortable repofe. This habitation is fo enviable, that the Fox, whofe abilities for burrowing in the ground are inferior to the Badger's, frequently ejects the lawful tenant by laying his fœtid excrement at the mouth of the hole; the Badger being fo cleanly an animal, that the calls of Nature are never obeyed within the apartment.

The female brings three or four cubs in fummer, which are fuckled for fome time, and afterwards are provided with fuch food as her abilities or induftry can procure. Badgers are frequently eaten, and are faid to make good bacon.

The hair of the Badger is very long, coarfe and rough, which gives it a very uncouth and clumfy appearance, and difguifes the true fhape of the limbs; each hair is tinged with three different colours; the roots a dirty white, the middle black, the extremity afh colour or grey, which has produced the well-known faying, "as grey as a Badger." The cheft and belly are very dark, nearly black; a ftripe of the fame alfo extends from the eye to the ear.

The Badger is common to moft northern countries, and are found in fome warmer ones: the Chinefe are very fond of their flefh, which is often an article in their butchers fhops; are hunted by night for the fake of the fkin; when attacked by the dogs, defend themfelves with great courage, and bite very feverely; are about two feet and a half long, tall about nine inches; have fmall eyes, and fhort round ears.

THE OTTER.

The O T T E R.

THE gradation from one class of beings to another is made by almost imperceptible degrees. The amphibious nature of some quadrupeds join them in close connection with the fish tribe; to which class, the links of the chain so gradually diminish, that we scarce know where to fix its termination. The Otter and Beaver are calculated to live on land, yet their propensity leads them to the water; and Nature has provided suitably for their destination. These animals have four feet, and the Beaver has a tail covered with scales; the gradation then descends to the Seal, whose hind feet answer more the appearance and purpose of fins; next the Walrus, &c. till all distinction is lost in a complete inhabitant of the limpid fluid.

The Otter is a very voracious animal, eats much, and destroys infinitely more; for, not content with sufficient to satisfy hunger, it kills through wantonness, and destroys for mere victory. On the brink of some lake or river, under the bank, where the waves have formed an excavation, in a gallery of this kind it makes its abode; and, when pursued, evades the search by plunging into the water many yards distant from the place where it was expected to be found. In a running stream the Otter always pursues its prey against the current; it frequently sinks to the bottom, and any fish passing over is sure to become its immediate prey. In standing water it hunts them into some creek, where they rarely escape its voracity or cruelty; for it will continue the hunt for hours, and dragging the prey on shore, leave it as trophies of success. In a few nights, one has been known to destroy all the fish in a large pond; will scent or wind the fish at a great distance.

The Otter, when taken young, may be tamed, and taught to follow like a dog, and may even be accustomed to fish for, and at the command of, his master: this perfection of education requires much perseverance, but is very profitable when attained. The Otter brings three or four young at a time; the old ones are rarely or never taken alive; the hunting them is esteemed good sport, as they sustain a long chace, fight boldly, and bite cruelly; indeed, few dogs will venture to attack them alone.

The colour of this animal is various; in general, a light-brown; neck, chest, and belly white; the hair rather coarse; neck long and thick; head round; eyes very close together; ears small; tail thickish; feet short, but very strong and flexible, and the articulations so loose, that they can be turned quite round, and brought on a line with the body; a membrane joins the toes of all the feet; about 2 feet long; tail 12 or 16 inches; are natives of most temperate parts of the world; and are found as far north as Kamschatka: the skins are esteemed very serviceable for gloves.

THE BEAVER.

THE BEAVER.

The B E A V E R

IS an amphibious animal; and, where the intrusion of man does not prevent, live together in a state of civil government, and appear the only influence of brutes forming a regular community, governed by domestic laws. The time of assembling is about the months of June or July, when a society is formed, which lasts the greater part of the year; the resort is from all quarters, and sometimes a troop of 200 or 300 assemble; the place of rendezvous is generally suitable for the colony, either on the banks of a lake, or on a running water. In the latter case, to guard against a sudden swell of the river, a bank or dam is formed across the stream, frequently of an hundred feet long; this is done by first driving flakes five or six feet long, placed on a row, with small twigs interwoven, and the interstices filled with clay; this dam is 10 or 12 feet thick at the base, and gradually diminishes to 2 or 3 at the top. The side next the head of water is sloped, the other is perpendicular.

The dam or mole being finished, the next care is to erect the several apartments or dwellings, which they build on piles or flakes drove into the ground for that purpose, and are either round or oval, divided into stories, to secure a retreat from swelling floods. The first is below the level of the dam, and is usually full of water; the walls are about 2 feet thick, made of earth, stones and sticks, most artfully laid together; the inside is neatly plaistered as with a trowel. Each house, which is about 8 feet above the water, has two openings, one into the water, the other towards the land. The size of the dwelling is proportioned to the number of Beavers which are to inhabit it, usually from 10 to 30. It has been observed, that 400 Beavers have resided in one large mansion, divided into a vast number of apartments, that had a free communication one with another. These works are finished by August or September; when they begin to lay in their stores, which consists principally of the wood of the birch, the plane, &c. which they steep in water, in quantities proper for use; the summer food is fresh leaves and fruits; are not fond of fish.

The benefits resulting from patient perseverance have become proverbial, and a more striking instance of the good effects cannot be given than the completion of these surprising works, which are begun by mere instinct, and are finished by mere industry. In the labours of this society every Beaver bears a part; some, by gnawing with their teeth, fell trees of great size, to serve as beams or piles; others drive them along the water, and, with their feet, scoop holes in which to place them; while others help to rear them up. Another party is employed in collecting twigs to weave between the flakes; a third in collecting earth, stones, and clay; while a fourth is busied in beating or tempering the mortar, which is done with the tail; others are employed in carrying it on the broad part of the tail to proper places, and with the same instrument ram it between the piles, or plaister the house. A certain number of smart strokes given with the tail, is a signal given by the overseer for repairing to certain places to mend any defect, or at the approach of some enemy; and the whole society attend with the readiest assiduity.

The teeth of the Beaver are admirably adapted for cutting timber, or stripping the bark, to which purposes they are so frequently applied. Is an inoffensive animal, and seeks safety rather in flight than conflict. The fur of the Beaver is of great service in the hat trade; it also produces a valuable drug, called castoreum; are hunted, and taken in traps and snares; inhabit most northern climes, but are no where found in such abundance as in Canada in America. The trade for Beavers furs with the Indians is a source of great wealth to the Hudson's Bay Company. The length of the Beaver is about 2 feet, height 1; tail 4 inches broad, 1 or 1¼ thick. The colour a fine chesnut brown; the hair of different lengths and fineness; is the only animal whose toes on the hind feet are joined by a membrane, while those on the fore feet are not; the front feet supply the place of hands, similar to the Squirrel.

THE GLUTTON

The G L U T T O N.

THIS singular animal, on account of the length of body, and short-ness of legs, appears to belong to the Weasel tribe. Mr. Pennant allots it with the Bear—it has a roundish head, with a blunt nose, short ears, limbs large and strong, tail very bushy, general colour black, with a broad horizontal stripe, of a yellowish colour, along the upper part of the face, the sides, and the tail.

The great voracity of this animal has fixed upon it the opprobrious name it bears. If the active speed which wild animals in general possess fell to the share of the Glutton, he must inevitably soon thin the forest of its inhabitants, but the cautious hand of Nature has guarded against his voracity by a body ill-formed for celerity; thus disqualified for pursuit, yet ever pressed by an active appetite, it has recourse to cunning and stratagem. Selecting a tree whose situation is promising, or observ-ing on the bark the marks of the teeth or horns of the deer, or other beast, he readily ascends, and, hiding among the spreading branches, he will there wait for weeks together, expecting some unwary animal to pass under, which he instantly drops upon, fixing his teeth and claws into the neck, digs a passage to the great blood vessels, which lie in those parts—in vain the tortured animal flies for relief among the branches of the forest, the Glutton still holds his station; and, although it often loses parts of its skin and flesh, which are rubbed off against the trees, yet it still flicks fast, the force of appetite and nature prevail more than his feel-ings; and he never seizes, but he brings down his prey, wearied by fatigue, and faint by loss of blood: the moment of victory rewards for former trouble, and he then makes up for past fatigue by imme-diately falling to, and ceases not, till overgorging has destroyed every animal function; thus torpid through satiety, he lies till nature qualifies him to renew the feast, which he does not quit till entirely eaten up bones and all. As such a bountiful repast cannot always supply his voracity, he uses much cunning to procure his prey; he will frequently anticipate the sportsman by clearing his traps of the game; he steals upon the retreat of other animals, particularly the rein deer, of whose flesh he is greedily fond: he also lies in wait, and falls upon the game other animals have run down, his constant necessities producing a pretty fertile invention. One of these animals confined at Dres-den consumed thirteen pounds of flesh every day, and yet not satisfied. The Glutton inhabits the northern parts of Europe, Siberia, and Ame-rica; its skin is highly esteemed for a beautiful gloss and damasked appearance; in length it is about three feet and a half, and eighteen or twenty inches high.

THE ARMADILLO.

The A R M A D I L L O.

WHEN we speak of a quadruped, imagination reprefents an animal covered with hair; as when we mention a bird, or a fifh, to the one we attribute feathers, to the other fcales; and thefe diftinctions, at the firft, appear to mark the boundary of each fpecies; yet nature, as if in defiance of rule, and wifhing to aftonifh as much by particular exceptions as by general laws, fo blends her feveral productions, that it is no eafy matter to draw a diftinguifhing line, and fay to which clafs an animal, whofe tail is covered with fcales, belongs; or of which family one inclofed in large fcales or fhells is a part. It therefore becomes us not to judge by one character only, which fo often is incomplete.

The Armadillo is one inftance among feveral of a quadruped covered not with hair, but with a fhell or fhells. Of this animal there are feveral kinds, whofe variety confifts in the number of the bands of fhell which encircle or cover them; to fome the incruftation is divided into only three diftinct pieces, to other into fix, eight, nine, twelve, and eighteen pieces; which have been confidered by fome as marks of age; but, in general, with more propriety, have been regarded as different kinds.

The bands of fhell lap over one another, and are united by a membrane, like the fhell to the tail of a lobfter; this fhell, or combination of fhells, covers the head, the upper part of the body, and the tail; the throat and belly being the only parts not fecured: this deficiency is provided againft by the power the Armadillo has of rolling itfelf up like a ball, and thereby covering the vulnerable parts. In time of danger, when it cannot make good its retreat to its hole, it brings the head and feet clofe to the belly, and, bending the back, forms nearly a fphere, the tail laps over the joining, and makes a firmer hold; in this form it defies the attack of any quadruped, and a patient fuffering of infult generally proves its fecurity; but man, whofe power is over the whole creation, whofe power and perfeverance is irrefiftible when any good is to be obtained, or any luxury enjoyed, foon convinces the poor Armadillo of its danger, by expofing it to the fire, which makes it quickly unroll.

The fmall kind of this animal are efteemed very nice eating, and are therefore hunted with avidity; dogs are ufed to purfue them, who impede their flight by making them roll themfelves up, when they become an eafy prey to man. As they run pretty faft, if a few minutes are allowed, they immediately fall to work, and feek fecurity by burrowing in the ground, which they do with great celerity, and muft then be dug out. Their accuftomed abode is in holes of confiderable depth, and, as they wander only by night, and then not far, fome induftry is required in fecuring them.

The colour of the fhell of the Armadillo is a greyifh yellow; that part of the head which is not covered, is a blackifh brown, the belly a-yellowifh white, which bears evident marks of a tendency to offify; the feet a flefhy red colour, are fpotted. Are natives of South America, particularly the Brafils; about 14 or 18 inches long; the larger kind 2 feet.

THE BOMBAY SQUIRREL.

The BOMBAY SQUIRREL.

THE tafk of the Hiftorian or Naturalift is often furrounded with perplexities and difficulties; fometimes, from the incertitude, the variation, and almoft total diffimilarity of his information; and fometimes from the entire want of every information; for, of thofe who are pleafed with poffeffing a new or ftrange animal, and will be at the trouble of tranfporting it from diftant parts, few have abilities or inclination, to feek that knowledge which would fatisfy a Zoologift in its habits and propenfities; and, perhaps, very few enjoy opportunities of acquiring certain intelligence.

With refpect to the animal now before us, we acknowledge our want of certain information; all we can fay is, that it was brought from India in one of the Company's fhips; it appeared to have all the motions and actions of a common Squirrel; and, as its fize was larger, fo its ftrength was greater. The fine rich colouring of the fur gave it a very grave and majeftic appearance. We are forry to add, that fince this drawing was made the animal is dead.

The length of the body was 15 inches; the tail as long; the head longifh and round; the ears tufted; the colour of the head and ears a fine deep brown; the fhoulders, along the back, hams and tail, black; fides a reddifh purple; cheft, fore feet, belly, and infide of hind feet, a yellowifh white. The end of the tail to this animal was not of an orange colour, as the one defcribed by Mr. Pennant was.

THE PECCARY.

Drawn from Nature Engraved & Publish'd by Cha.ton Catton Jun.r   Aug.t in the Terrace Hampton Court Road Oct.r 1 1788.

The P E C C A R Y,

AT first view, bears a general resemblance to our common hog, but on examination is evidently of a distinct species; neither will they breed together. The head of the Peccary is large; the snout long, and terminates like the hog's; the neck is thick and short; the body bulky, and marked down the neck with a belt of a whitish colour; the legs are short; the general colour is black; each hair or bristle is marked with alternate bands of black and white, like the porcupine's quills; the coating is a coarse kind of bristles, which are long over the whole body, and the length of four or five inches along the back; has no tail; the face is rather smaller than the common hog; the appearance equally clumsy with all of this tribe.

The Peccary is further distinguishable from every other quadruped, by an orifice in the back, near the rump, which by some has been mistaken for the navel; from this opening discharges an ichorous liquor of a disagreeable smell. It is necessary, immediately on killing the animal, to extract this orifice or gut, else, in the course of a quarter of an hour, it will taint the whole carcase.

The Peccary is a native of the hottest parts of South America, where they are very numerous, and go in, herds of two or three hundred; prefer the mountains to the plains, and the woods to the open parts, as the food they most delight in abounds there in the greatest plenty; they eat also toads, lizards and serpents; the latter they skin with great adroitness, holding them with the fore feet.

The Peccary, though not armed with such offensive weapons as the wild boar or hog, will fight stoutly with the beasts of prey. The Jaguar, or American Leopard, is its mortal enemy; often the body of that animal is found with several of these hogs, slain in combat. They render dear mutual assistance when attacked, and endeavour to surround the enemy. The Peccary may be rendered tame and domestic; is satisfied with the same food as the hog, but is not so much inclined to be fat; nor will it, like them, wallow in mire. The flesh is esteemed very good food.

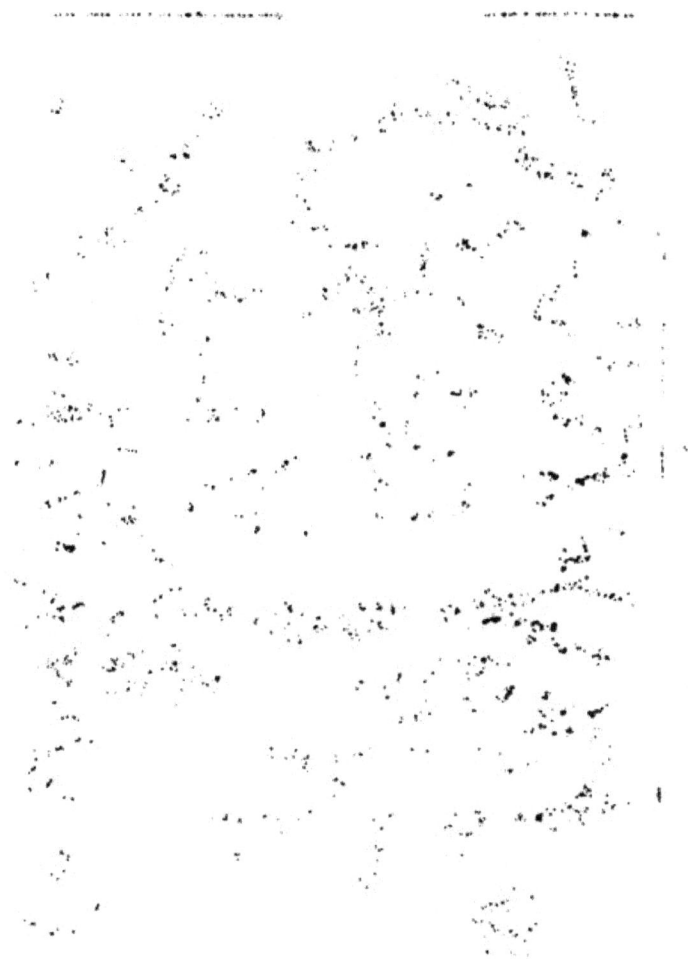

The P O R C U P I N E.

THE very fingular properties of this animal have been the foundation of many fabulous reports.

Nature, in all her productions, gives to every animal fome particular quality—in fome fhe implants a fierce and favage cruelty, regarding only the gratification of an inordinate appetite; in others a mind harmlefs and peaceable, yet poffeffed of powers and faculties to act on the defenfive, fuch as extreme caution, or cunning, which may be called an internal defence; or an external one, fuch as fhields the animal now before us—thus guarding the weaker againft the overbearing oppreffion of the ftronger, thereby preferving an equality and balance in her productions. The Porcupine is in its nature quiet, and feldom gives provocation of offence; and when attacked by an animal of prey defends itfelf by erecting its quills in fuch a manner as always to keep them pointed towards the enemy; thus fecuring its own fafety. Sparrman, a modern traveller to the Cape, reports, "By rolling up its body like the "hedge-hog into a heap, and fetting up its prickles, or quills, many of "which are a foot and a half long, it is perfectly well defended from "dogs, as well as other animals."—If time permits, it makes towards, and afcends a tree, where perched in fafety, it wearies the patience of its purfuer.

The power formerly attributed to the porcupine of voluntarily difcharging its quills, and with them mortally wounding, at a confiderable diftance, is now entirely difcredited, great provocation having been ufed, if poffible, to produce this effect, but without fuccefs. The quills are ftrongly inferted in the body of the animal; each one fed or fupported by a fmall ball or nucleus of a foft fpongy matter, varying in fize according to the bulk of the quill.

Thofe Porcupines which inhabit the Eaft are reported to poffefs a Bezoar or ftome, which is reputed an antidote to poifon; this is found in the head, and is confidered of great value. Taverner reports his giving five-hundred crowns for one, which he afterwards changed to advantage; he fays they are alfo found in the belly fometimes. Of this animal there are feveral fpecies little akin otherwife than in being provided with a coat of quills. The one here drawn has the upper lip divided, head like an hare, with a row or ruff of ftiff briftles furrounding it on the forepart of the fhoulders and top of the head, reclining backwards; the body part is thickly covered with quills from nine to twelve inches long, very fharp at the point, and regularly annulated with alternate black and white; fome of the larger quills are near a quarter of an inch diameter; the internal fubftance is fpongy, like the upper part of a goofe-quill; the body is thickly covered with hair between the quills; the head, belly, and legs are covered with ftrong briftles rooted among foft hair; the feet are fhort, as is the tail, which is covered with quills; the general length of the animal is about three feet; it inhabits Africa, India, Tartary, and Paleftine. In Italy a fpecies with fhorter quills run wild; thefe are fold in the market at Rome, where they are eat:—The traveller before quoted fays, "the flefh neareft re-"fembles pork, a circumftance which undoubtedly gave it the name it "bears; it is chiefly ufed as bacon, being fmoked and dried up the "chimney for that purpofe, and is by no means ill-tafted." It feeds on fruits, roots, and herbs; the colour inclines to black.

HIPPOPOTAMUS.

Drawn from the Original by J.Ireland & Pubd. by Revd. Cotton & for. S.t. John's church.

S.t. Pauls Church-Yard London Event Pubd. Aug.t 10 1787.

The HIPPOPOTAMUS.

THIS is an amphibious animal, of a gigantic bulk, inferior only to the Elephant; is found in large and rapid rivers; the Nile, the Gambia, and the great rivers of the South-East parts of Africa. The general dimensions of this animal are reported as follows; in length about 17 feet, circumference 15, height 7, legs 3, head 3½ feet long, and 9 round; tail shortish, eyes and ears small; the latter pointed, covered on the inside with hair; the two tusks in the lower jaw lie nearly horizontal, and measure about 27 inches long; their weight about 3½ pounds; these tusks are highly prized by Dentists, for making false teeth, not being so liable to turn yellow as ivory. The skin thinly interspersed with strong hairs, or bristles; is very rough, of a mud colour, and, when newly risen out of the water, the animal has a glittering or slimy appearance.

This animal, unless insulted, is of a quiet disposition; passing its time in wandering up and down the rivers, which it chiefly inhabits. Fish, as some have reported, makes no part of its food; this it seeks at night on land, feeding on the grass, reeds, and boughs of trees which are in the neighbourhood of its abode. "The quantity of grass," says Mr. Sparman, "which I have at different times observed to have been confumed "by one of these animals, in spots where it has come over night to "graze, is almost incredible." Indeed, considering the bulk of its body, and the great size of its stomachs (which are four), it certainly must require great quantity of nutriment. When in the water, the Hippopotamus frequently rises to the top to take in air; it will suffer immersion about 30 or 40 minutes. Providence, in its universal wisdom, has appointed the abode of this colossal animal in very distant parts from the habitations of man, else its great strength and revengeful nature must produce much mischief;——of its strength, it has the credit of being able, with ease, to bite a man in two; and of its revengeful nature, it has been known to place itself under a boat, and by rising up, overset it with six men in. Moore, in his travels up the Gambia, relates a similar disaster: A boat going down the river, fell in with a herd of these animals: "on being fired at," says the Narrator, "before the flashing of "the pan was well out of our eyes, being in the midst of them, one "which we supposed was wounded, flounced and kicked about the "boat till he knocked a piece out of the bottom; and before we could "reach shore, she sunk right down." One of these animals pursued for several hours a Hottentot, who found it very difficult to make his escape; their activity when on land must not, therefore, be calculated from the unwieldiness of their appearance. In the water they swim with great vigour against the strongest current, and frequently sink to the bottom, and walk as on land: they frequent salt water, but do not drink it; when angry, make a furious noise, between the grunt of a hog and the neighing of a horse, which, probably, fixed on it the name of Hippopotamus, which is Greek, and signifies a river horse; its number of stomachs has certainly caused it to be called the sea cow.

A considerable portion of a skeleton of one was lately found in digging at Chatham, which has led a learned Gentleman to make some ingenious queries concerning the antiquity of the earth, the climate of this country in former times, and to conclude this animal was once a native of England. The Hippopotamus is certainly the animal which is described in such a figurative, yet correct, manner in the 40th chapter of Job.

The drawing for this subject was taken from a stuffed skin in the Leverian Museum, and is regarded as a just figure of the animal, though not at full growth; the dimensions were 9 feet long, and 5½ high.

CROCODILE.

Drawn from Life Engraved & Published by Col.ᵉˡ Cotton Junᵉ for it, Act done.

Nᵒ on the Terrace Tavistock Covent Garden Aug.ᵗ 14 1.

The CROCODILE.

THIS amphibious animal is claffed among the Lizard Tribe. A courage fierce and favage, aided by great bodily ftrength, joined to a confiderable fhare of cunning, or ftratagem, compofe the great outline of this animal's character.

Of a bulk truly formidable from 18 to 28 feet long, they are univerfally dreaded; always on the watch, with activity and appetite ever ready, the Crocodile lets flip no opportunity of committing his depredations on animal nature; the water is his proper element, but if his voracity has caufed a fcarcity of game here, hid among the reeds, he lies in wait on the banks of the river, expecting the approach of fome thirfty animal, compelled by the heat to regale nature with a lap of water, then the Crocodile immediately feizes upon, and pulls down his prey; where, unlefs of very large bulk, it rarely efcapes being prefently drowned, holding his prey both by his claws and his mouth, which, in one 17 feet long, will open near 2 feet, a gape fufficient to take firm hold of man or beaft.

This animal is oviparous, or generated by eggs, which the female depofits with the utmoft fecrecy and circumfpection in the fand, on the fhore of the river; fcratching a hole in a fuitable place, in about an hour fhe depofits near an hundred eggs, then covers the place with the moft fedulous anxiety for their fafety; the fame tafk is performed the fucceeding and third day, when about 300 eggs are depofited, thefe covered with great care with the fand, fhe commits to the foftering hand of Nature: the heat of the fun in about 30 days animates the eggs, and now Nature prompts the mother to feek after her young by clearing away the fand; the brood thus liberated, fome take immediately to the water, while others mounted on her back, are introduced to their fluid habitation with more eafe and fafety; this parental care foon fubfides; their proper element once gained, fafety depends on their agility and caution.

The ferocity of the Crocodile, like other wild animals, very much abates as his abode is more or lefs in an inhabited country. In unfrequented rivers they lie bafking in droves together, and have the appearance of large trunks of trees, with rough and rugged bark floating on the water; yet, thus apparently torpid, appetite is awake, and the approach of any animal is quickly followed by a conflict for victory. In more populous countries the undivided tyranny of man has reduced them within better bounds. It is reported, the children of the Siamefe play with them in a very familiar manner, and will even correct them with blows; it is true, indeed, thefe people treat them more as friends than enemies. The reverfe to this is the general character of the Crocodile, whofe great fecundity muft be very alarming, had not the wife, the beneficent hand of Providence appointed a bird of the Vulture kind, and an animal called the Ichneumon, with an appetite peculiarly fond of the Crocodile's eggs.

The colour is a greenifh brown, the upper part of the body is covered with a very thick and rough fkin, proof againft the edge of a fword; the belly, of a greenifh white, is more vulnerable; the eye is very prominent and large, of a yellowifh green; the feet are fhort, but very mufcular; with its tail it ufually knocks down and ftuns its prey. Inhabits moft great rivers of Afia, Africa, and America: the Nile in Egypt has ever been famous for them.

www.ingramcontent.com/pod-product-compliance
Lightning Source LLC
Chambersburg PA
CBHW021703210326
41599CB00013B/1501